tech excuses for beginners

By The Employee of the Month

Copyright © 2021
The Employee of the Month
All rights reserved.
ISBN: 9798458529402

This book is for anyone who is looking for a solution to their programming problems.

Learn fun and alternatives ways for solving problems that a developer encounters every day.

HOWTO

1. Choose the page side (left or right)
2. Close your eyes and think about your problem
3. Open the book and read the solution

Obviously, the problem will continue to persist :) but at least you have taken a break from this intense research and who knows it will help you come out with some workaround.

I hope it will bring some smiles to you in the everyday work in your Agile teams.

You can also add your own answers to make this book more personal for you and your team.

Enjoy!

Maybe

Maybe _____

Maybe _____

Maybe _____

Maybe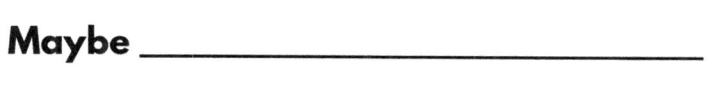

Maybe _____

Maybe no one understand me

Maybe this is not the job for me

Maybe try with another user

Maybe it's still in the browser cache

Maybe you should check the DB connection

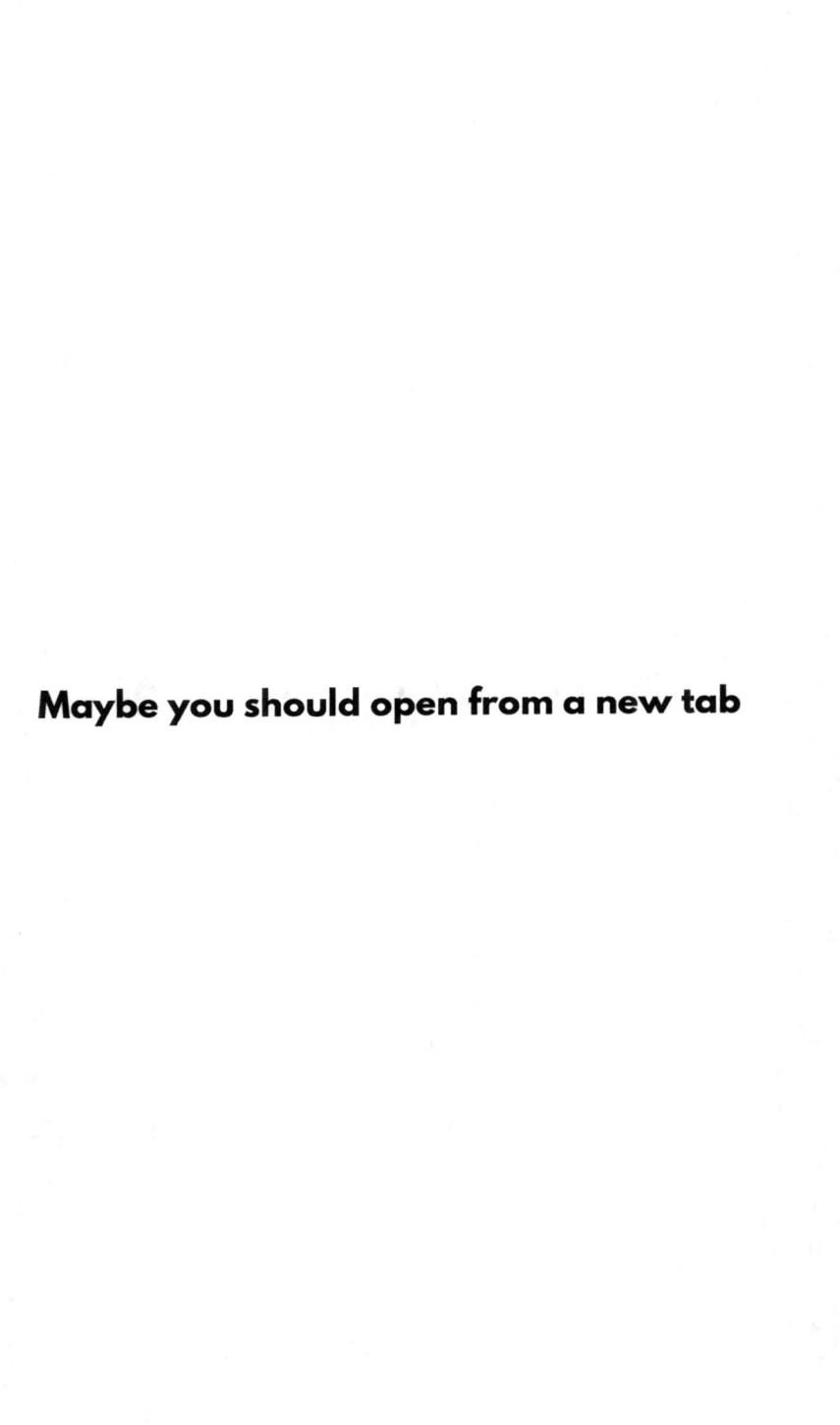

Maybe you should delete cookies

Maybe I should find a workaround

Maybe I should find a hotfix

Maybe I should ask to the architect

Maybe it's the proxy

Maybe it's the firewall

Maybe your system version does not support it

Maybe you have too many tabs open

Maybe you have too many applications open

Maybe it's not released yet

Maybe it's not deployed

Maybe I have not pushed

Maybe try later

Maybe Monday it will work

Maybe tomorrow it will work

Maybe it's not responsive

Maybe I will search on Google

Maybe it's too much for today

Maybe your not admin

Maybe your user it's not authorised

Maybe I have not understood the specification

Maybe you are right

Maybe I will post a question on StackOverflow

Maybe I should wait

Maybe I should go on holiday

Maybe I should not have copied the answer from StackOverflow

Maybe I should search on StackOverflow

Maybe it's the hosting

Maybe you should change the resolution

Maybe I should downgrade the framework

Maybe I should upgrade the framework

Maybe it's in DEV

Maybe it's not in production

Maybe it's not Agile

Maybe it's not mobile first

Maybe it's happening only on retina display

Maybe it's the WIFI

Maybe it's the wrong version

Maybe I should upgrade the environment

Maybe I should restart the computer

Maybe it's the VPN connection

Maybe you are too fast

Maybe it's the codec

Maybe I should refresh

Maybe you should change browser

Maybe I need to work peacefully without disturbance

Maybe I should restart the server

Maybe there are too many microservices

Maybe you have run out of licenses

Maybe I should go on the cloud

Maybe vi needs to be upgraded to vii

Maybe I'm too old for that

Maybe there is not enough RAM

Maybe it's the millennium bug

Maybe the async call is too lazy

Maybe it's in recovery mode

Maybe it's the big to little endian conversion

Maybe the monitor is plugged in the serial port

Maybe it's Monday

Maybe the connection is refused

Maybe daemons loose in system

Maybe it's your machine

Maybe it's the cache

Maybe processes running slowly due to weak power supply

Maybe it's the Antivirus

Maybe it's not plugged in

Maybe it's the DNS

Maybe it's the virtual machine

Maybe the response is empty

Maybe it's the OS

Maybe there is a bug in the RAID

Maybe I should come back to my mum house

Maybe I should take a break

Maybe it does not escalate

Maybe it does not exist

Maybe it's the intenet connection

Maybe it's not you, it's me

Maybe it's not working

Maybe you should try later

Maybe it's my fault

Maybe I should fix it

Maybe there are too many requests

Maybe I should take some time for myself

Maybe you don't understand me

Maybe it's the API

Maybe it's a bad user karma

Maybe you should login

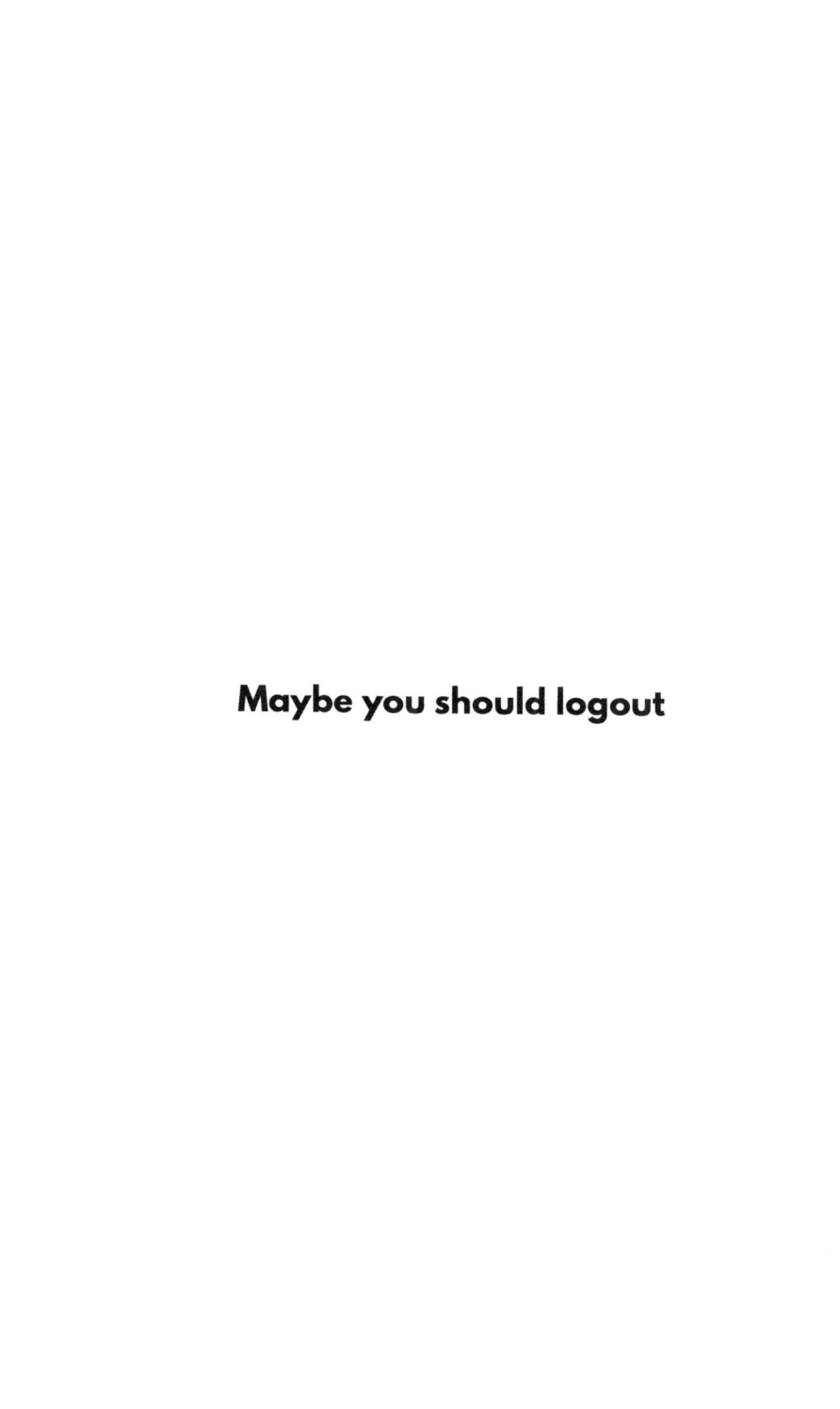

Maybe it's not working only for that user

Maybe it's the software upgrade

Maybe it's the internet connection

www.ingramcontent.com/pod-product-compliance
Lightning Source LLC
Chambersburg PA
CBHW052329220526
45472CB00001B/332